YOUR KNOWLEDGE HAS VALUE

- We will publish your bachelor's and
  master's thesis, essays and papers

- Your own eBook and book -
  sold worldwide in all relevant shops

- Earn money with each sale

Upload your text at www.GRIN.com
and publish for free

Albert H. Kaiser

Digital Signal Processing using the Fast Fourier Transform (FFT)

GRIN Verlag

Bibliografische Information der Deutschen Nationalbibliothek:

Die Deutsche Bibliothek verzeichnet diese Publikation in der Deutschen National-
bibliografie; detaillierte bibliografische Daten sind im Internet über http://dnb.d-
nb.de/ abrufbar.

Dieses Werk sowie alle darin enthaltenen einzelnen Beiträge und Abbildungen
sind urheberrechtlich geschützt. Jede Verwertung, die nicht ausdrücklich vom
Urheberrechtsschutz zugelassen ist, bedarf der vorherigen Zustimmung des Verla-
ges. Das gilt insbesondere für Vervielfältigungen, Bearbeitungen, Übersetzungen,
Mikroverfilmungen, Auswertungen durch Datenbanken und für die Einspeicherung
und Verarbeitung in elektronische Systeme. Alle Rechte, auch die des auszugsweisen
Nachdrucks, der fotomechanischen Wiedergabe (einschließlich Mikrokopie) sowie
der Auswertung durch Datenbanken oder ähnliche Einrichtungen, vorbehalten.

Imprint:

Copyright © 1997 GRIN Verlag GmbH
Druck und Bindung: Books on Demand GmbH, Norderstedt Germany
ISBN: 978-3-638-63914-9

This book at GRIN:

http://www.grin.com/en/e-book/5978/digital-signal-processing-using-the-fast-fou-
rier-transform-fft

GRIN - Your knowledge has value

Der GRIN Verlag publiziert seit 1998 wissenschaftliche Arbeiten von Studenten, Hochschullehrern und anderen Akademikern als eBook und gedrucktes Buch. Die Verlagswebsite www.grin.com ist die ideale Plattform zur Veröffentlichung von Hausarbeiten, Abschlussarbeiten, wissenschaftlichen Aufsätzen, Dissertationen und Fachbüchern.

Visit us on the internet:

http://www.grin.com/

http://www.facebook.com/grincom

http://www.twitter.com/grin_com

Department of Aeronautical and Automotive Engineering and Transport Studies

MSc in Automotive Systems Engineering

Module 1: Engineering Framework

Course :

Signal Analysis

Lecturer:

Dr. S. Walsh

G. Jenkins

Coursework :

Digital Signal Processing using the Fast Fourier Transform

by:

Albert Kaiser

date: June 1997

1

Summary

Conventionally a *signal* is a physical variable that changes with time and contains information. The signal may be represented in *analogue* (continuos) or *discrete* (digital) form. The majority of the physical variables of interest for the engineer are of analogue form. However digital data acquisition equipment favour a digital representation of the analogue signal.

The digital representation of a analogue signal will effect the characteristic of the signal. Thus an understanding of the underlying principles involved in signal processing is essential in order to retain the basic information of the original signal.

The primary goal to use the Discrete Fourier Transform (DFT) is to approximate the Fourier Transform of a continuous time signal. The DFT is discrete in time and frequency domain and has two important properties:

- the DFT is periodic with the sampling frequency
- the DFT is symmetric about the Nyquist frequency

Due to the limitations of the DFT there are three possible phenomena that could result in errors between computed and desired transform.

- Aliasing
- Picket Fence Effect
- Leakage

The DFT of a signal uses only a finite record length of the signal. Thus the input signal for the DFT can be considered as the result of multiplying the signal with a window function. Multiplication in the time domain results in convolution in the frequency domain, which will influence the spectral characteristic of the sampled signal. In the table below rectangular and Hanning window are compared:

		Rectangular Window	Hanning Widow
peak value (input: c_n)	-	$c_n \cdot T$	$c_n \cdot T \big/ 2$
fist zero crossing frequency	[Hz]	$1 \big/ T$	$2 \big/ T$
attenuation	[dB]	-20	-60
'smearing'	-	small	large
side lobes	-	large	small
leakage	-	yes	no

The Fast Fourier Transform (FFT) is a computationally efficient algorithm for evaluating the DFT of a signal. It is imported to appreciate the properties of the FFT if it is to be used effectively for the analysis of signals. In order to avoid aliasing and resulting misinterpretation of measurement data the following steps should be followed:

1. decide on frequency range of interest, e.g. the bandwidth.
2. match the bandwidth of the anti aliasing filter, e.g. a low pass filter with cut off frequency set to bandwidth
3. choose sampling frequency f_s of analogue digital converter to be 2.5 to 5.0 times the cut of frequency
4. choose window length to give required frequency resolution.

$$T = 1 \big/ f_{fr}$$

5. check number of samples $N = T \cdot f_s$, for efficient use of the FFT it is required that:

$$N = 2^m$$

6. repeat step 3, 4 and 5 until an acceptable compromise if found

The effect of random noise can be minimised by averaging several spectra of a signal.

Table of Contents

Notation

T	period	[s]
T_0	record length	
N	number of samples	
τ	sampling interval	
n, m	integer counters	
a_n, b_n, c_n	coefficients of Fourier Series	
f	frequency	[Hz]
f_{sr}	spectral resolution	
f_{fr}	frequency resolution	
f_s	sampling frequency	
f_{ny}	Nyqist frequency	
ω_0	fundamental frequency (angular velocity)	[rad / s]

Abbreviations

FS	Fourier Series
FT	Fourier Transform
DFT	Discrete Fourier Transform
IDFT	Inverse Discrete Fourier Transform
FFT	Fast Fourier Transform
IFFT	Inverse Fast Fourier Transform

Note: Throughout the text angular velocity and frequency are used interchangeably, since both are found in literature. They are related with the simple formula $\omega = 2\pi \cdot f$. Technical publications favour frequency.

1 Objective

To use MATLAB to demonstrate the properties of the Fast Fourier Transform (FFT) used for the spectral analysis of signals.

2 Approach

- A theoretical background of the basic principles of digital signals processing is given in Chapter 4.
- MATLAB was used to demonstrate the properties of the Fast Fourier Transform. The FFT of known input signals were investigated (Chapter 5).

3 Introduction

Acquisition of data is of vital importance in automotive industry. Development of a vehicle, as well as modern on board electronic systems relay on sophisticated data acquisition systems.

In the development process the ultimate target is to translate customer wants into objective engineering targets. At FORD the Quality Function Deployment (QFD) process is used. Traditionally the vehicle attributes are divided into the categorises Noise Vibration Harshness (NVH), Vehicle Dynamics (VD), Safety and Durability.

A few examples for data collected are:
- Noise Vibration Harshness: Point Mobility, Noise Transfer Functions, interior noise,
- Vehicle Dynamics: ride and handling, ride comfort
- Crash: decelerations, peak dummy forces
- Durability: road load data

Modern on board electronic systems that relay on digital data acquisition are for example:
- motor management systems
- antilock braking systems (ABS)
- airbag
- traction control and acceleration slip regulation (ASR)
- active suspension system

The underlying principle for all these measurements is identical. A physical quantity is measured with a sensor that produces a analogue signal. This analogue signal is changed into digital form using an analogue digital converter. The digital signal is then processed using digital equipment (computers) and recorded for later evaluation. In the case of the onboard systems the digital signal is used for digital control of the system. The output of the controller is then feed into a digital analogue converter that drives an actuator.

The variety of applications of digital signals processing emphasises the importance of an understanding of digital signal processing and the underlying principles.

4 Analytical

4.1 Introduction

Conventionally a**signal** is a physical variable that changes with time and contains information. The signal may be represented in**analogue** (continuos) or**discrete** (digital) form. The majority of the physical variables of interest for the engineer are of analogue form. However digital data acquisition equipment favour a digital representation of the analogue signal.

The digital representation of a analogue signal will effect the characteristic of the signal. Thus an understanding of the underlying principles involved in signal processing is essential in order to retain the basic information of the original signal.

4.1.1 Classification of Signals

A large amount of signals can be classified as**deterministic**, i.e., mathematical expressions can be written which will determine their instantaneous value at any time. The deterministic signal may be divided into**periodic** signals, e.g., signals that repeat itself at a constant interval of time and **non periodic** signals. *Transient* and *almost periodic* signals fall in the later class.

There are, however signals that are**random**, e.g., further instantaneous values of the signal can not be predicted in a deterministic sense. For a*stationary* random signal the statistical properties (mean , mean square, variance, standard deviation) of the random signal do not change with time. Otherwise the signal is called*non stationary.*

4.1.2 Periodic Signals

A signal$f(t)$ is said to be periodic if it repeats itself in constant intervals of time. The repetition time is called period. It is obvious that integer multiples of the period are also a period of the signal. The smallest period is called the*primitive period T* , in technical literature the primitive period is often called period.

$$f(t) = f(t+nT) \qquad \text{where: n = integer} \qquad (1)$$

4.2 Fourier Series

The French mathematician J. Fourier (1768-1830) showed that any periodic function can be represented by a series of sins and cosines which are harmonically related. If$f(t)$ is a periodic function of period$T$, it is represented by the Fourier Series (the proof may be found in basic text books on calculus[1]):

$$f(t) = \frac{a_0}{2} + \sum_{n=1}^{\infty} a_n \cos(n\omega_0 t) + \sum_{n=1}^{\infty} b_n \sin(n\omega_0 t) \qquad (2)$$

with the fundamental frequency: $\qquad \omega_0 = \frac{2\pi}{T} \qquad n = 1, 2, 3, ..$

The Fourier Coefficients are:

$$a_n = \frac{2}{T} \int_{-T/2}^{T/2} f(t)\cos(n\omega_0 t)dt \qquad (3)$$

$$b_n = \frac{2}{T} \int_{-T/2}^{T/2} f(t)\sin(n\omega_0 t)dt \qquad (4)$$

---

[1] E. Kreysing, page 465-467

The Fourier Series can be expressed in complex form[2]:

$$f(t) = \sum_{n=-\infty}^{\infty} c_n e^{in\omega_0 t} dt \qquad (5)$$

where:

$$c_n = \frac{1}{T} \int_{-T/2}^{T/2} f(t) e^{-in\omega_0 t} dt \qquad (6)$$

Note that a continuous periodic function $f(t)$ in the time domain is represented by a discrete spectrum in the frequency domain, at frequencies which are integer multiples of the fundamental frequency.

4.2.1 Fourier Series of a square wave

The square wave is defined by:

$$f(t) = \begin{cases} 1, & 0 \le t \le T/2 \\ 0, & T/2 < t < T \end{cases} \qquad (7)$$

The Fourier Coefficients are obtained from:

$$c_n = \frac{1}{T} \int_0^{T/2} 1 \cdot e^{-in\omega_0 t} dt$$

The dc component follows for n=0:

$$c_0 = \frac{1}{T} \int_0^{T/2} 1 \cdot dt$$

$$c_0 = \frac{1}{2}$$

The other coefficients are:

$$c_n = \frac{1}{T} \left[\frac{-e^{in\omega_0 t}}{i \cdot n\omega_0} \right]_0^{T/2}$$

substituting the limits and with: $\omega_0 = \dfrac{2\pi}{T}$

$$c_n = \frac{-e^{-in\pi} + 1}{i \cdot n2\pi}$$

$$c_n = \frac{1}{n\pi} \cdot e^{-in\frac{\pi}{2}} \cdot \frac{e^{in\frac{\pi}{2}} - e^{-in\frac{\pi}{2}}}{2 \cdot i}$$

$$c_n = e^{-in\frac{\pi}{2}} \cdot \left(\frac{1}{n\pi} \cdot \sin\left(\frac{n\pi}{2}\right) \right) \qquad \text{for } n = \pm1, \pm2, \dots, \pm\infty$$

$$c_n = e^{-in\frac{\pi}{2}} \cdot (c_{n'}) \qquad (8)$$

where:

$$c_{n'} = \left(\frac{1}{n\pi} \cdot \sin\left(\frac{n\pi}{2}\right) \right) \qquad (9)$$

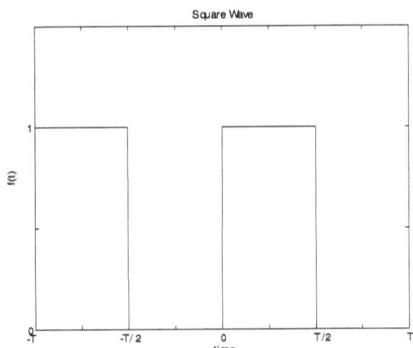

- figure 1: square wave

---

[2] W. Thomson, page 7

The term $e^{-i \cdot n \frac{\pi}{2}}$ in the equation (8) represents a phase shift. The spectrum for a square wave is given by equation (9):

$$c_{0'} = \frac{1}{2}$$

$$c_{1'} = \frac{1}{\pi}, \qquad c_{2'} = 0, \qquad c_{3'} = -\frac{1}{3\pi}, \qquad , \dots$$

$$c_{-1'} = \frac{1}{\pi}, \qquad c_{-2'} = 0, \qquad c_{-3'} = -\frac{1}{3\pi}, \qquad , \dots$$

Substituting the coefficients in the Fourier Series gives:

$$f(t) = \sum_{n=-\infty}^{+\infty} c_n \cdot e^{i \cdot n \omega_0 \cdot t}$$

$$f(t) = \sum_{n=-\infty}^{+\infty} e^{-i \cdot n \frac{\pi}{2}} \cdot c_{n'} \cdot e^{i \cdot n \omega_0 \cdot t}$$

The phase shift was: $\qquad t_0 = \frac{T}{4} = \frac{2 \cdot \pi}{\omega_0 \cdot 4} \qquad$ thus: $\qquad \frac{\pi}{2} = \omega_0 \cdot t_0$

After simplification the above equation becomes:

$$f(t) = \sum_{n=-\infty}^{+\infty} c_{n'} \cdot e^{i \cdot n \omega_0 \cdot (t - t_0)} \tag{10}$$

$$f(t) = g(t - t_0)$$

Thus the Fourier Series of a function $f(t)$ which has a phase shift with respect to $g(t)$ may directly be obtained from $g(t)$. The phase shift should be chosen to facilitate the evaluation of Fourier coefficients.

4.3 Fourier Transform

4.3.1 Definition of the Fourier Transform

The discrete spectrum of a Fourier Series becomes continuous when the period T is extended to infinity. The Fourier Transform can be regarded as the limiting case of the Fourier Series as the period approaches infinity.

The Fourier Transform pair is given by[3]:

Fourier Integral: $\qquad f(t) = \frac{1}{2\pi} \int_{-\infty}^{\infty} F(\omega) e^{i\omega t} d\omega \qquad$ (11)

Fourier Transform: $\qquad F(\omega) = \int_{-\infty}^{\infty} f(t) e^{-i\omega t} dt \qquad$ (12)

or in terms of frequency:

Fourier Integral: $\qquad f(t) = \int_{-\infty}^{\infty} F(f) e^{i2\pi f t} df$

Fourier Transform: $\qquad F(f) = \int_{-\infty}^{\infty} f(t) e^{-i2\pi f t} dt$

Note that a continuos function in the time domain $f(t)$ is resolved into a continuos spectrum in the frequency domain.

---

[3] W. Thomson, page 421

4.3.2 Fourier Transform of a sin wave

The sin wave may be written as:

$$f(t) = b_n \sin(2\pi f_n t)$$

or in complex form:

$$f(t) = b_n \cdot \frac{e^{i \cdot 2\pi f_n t} - e^{-i \cdot 2\pi f_n t}}{2 \cdot i} \qquad (13)$$

$$f(t) = \frac{-i \cdot b_n}{2} \left(e^{i \cdot 2\pi f_n t} - e^{-i \cdot 2\pi f_n t} \right)$$

Consider the simple periodic function:

$$x(t) = B \cdot e^{i 2\pi f_n t} \qquad (14)$$

Substituting in the Fourier Integral:

$$B \cdot e^{i 2\pi f_n t} = \int_{-\infty}^{+\infty} X(f) \cdot e^{i 2\pi f \cdot t} df$$

With the dirac function (unit impulse function) the above equation is satisfied

$$X(f) = B \cdot \delta(f - f_n) \qquad (15)$$

Using this result for the original problem gives immediately:

$$F(f) = \frac{-i \cdot b_n}{2} \left[\delta(f - f_n) - \delta(f + f_n) \right] \qquad (16)$$

Thus the Fourier Transform of a sine wave with frequency f_n is given by two discrete spectral lines at frequency $\mp f_n$ with amplitude $\pm \frac{b_n}{2}$.

4.3.3 Fourier Transform of a square wave

The square wave was defined by (7):

$$f(t) = \begin{cases} 1, & 0 \le t \le T/2 \\ 0, & T/2 < t < T \end{cases}$$

This may be expressed in terms of a Fourier Series (10):

$$f(t) = \sum_{n=-\infty}^{+\infty} c_{n'} \cdot e^{i \cdot n\omega_0 \cdot (t - t_0)}$$

the Fourier Coefficient are (9):

$$c_{n'} = \left(\frac{1}{n\pi} \cdot \sin\left(\frac{n\pi}{2} \right) \right) \qquad \text{for } n = \pm 1, \pm 2, \ldots, \pm\infty$$

$$c_{0'} = \frac{1}{2}$$

the phase shift was: $\quad t_0 = \dfrac{T}{4}$

with: $\quad f_n = \dfrac{n \cdot \omega_0}{2\pi}$

The FT of the Fourier Series defined by equation (10) is obvious considering each individual term in the Fourier Series and recognising the similarity to equation (14):

$$\therefore \quad F(f) = \sum_{n=-\infty}^{\infty} c_{n'} \cdot \delta(f - f_n) \qquad (17)$$

4.3.4 Fourier Transform of a rectangular pulse

A single rectangular pulse is defined by:

$$f(t) = \begin{cases} 1, & -T/2 \le t \le T/2 \\ 0, & \pm\infty < t < \pm T/2 \end{cases}$$

(18)

Substituting in the Fourier Transform gives:

$$F(f) = \int_{-T/2}^{T/2} 1 \cdot e^{-i \cdot 2\pi f \cdot t} dt$$

$$F(f) = -\frac{1}{i \cdot 2\pi f} \cdot \left[e^{-i \cdot 2\pi f \cdot t} \right]_{-T/2}^{T/2}$$

$$F(f) = -\frac{1}{i \cdot 2\pi f} \cdot \left[e^{-i \cdot \pi fT} - e^{i \cdot \pi fT} \right]$$

$$F(f) = \frac{T}{\pi fT} \cdot \left[\frac{e^{i \cdot \pi fT} - e^{-i \cdot \pi fT}}{2i} \right]$$

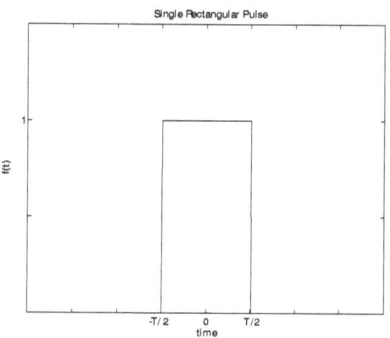

- figure 2: single rectangular pulse

$$F(f) = T \cdot \frac{\sin(\pi T \cdot f)}{\pi T \cdot f}$$

(19)

$$F(f) = T \cdot \mathrm{sinc}(\pi T \cdot f)$$

The resulting spectrum is a continuous function of the frequency. The cross over frequency of the spectrum are found at $f = \frac{n}{T}$ (n = integer). At these frequencies the pulse contains no energy. For a long duration of the pulse (T large) the spectral energy is concentrated around frequency zero (f=0). If the duration of the pulse is short (T small) than the spectrum is flat and extends throughout the frequency band.

4.3.5 Window Functions

The discrete Fourier transform of a signal uses only a finite record length of the signal. Thus the input signal for the DFT can be considered as the result of multiplying the signal with a window function. The most basic window function is the rectangular window.

The windowing process may be written mathematically as:

$$z(t) = x(t) \cdot y(t)$$

(20)

with:
- actual signal: $x(t)$
- window function: $y(t)$
- recorded signal: $z(t)$

4.3.5.1 Convolution in Frequency Domain

The Fourier Transform of the recorded signal is:

$$Z(\omega) = \int_{-\infty}^{\infty} x(t) \cdot y(t) \cdot e^{-i \cdot \omega \cdot t} dt$$

For a window function $y(t)$ the Fourier transform is defined by the pair:

$$y(t) = \frac{1}{2\pi} \int_{-\infty}^{\infty} Y(\omega') \cdot e^{-i\omega' t} d\omega' \qquad Y(\omega') = \int_{-\infty}^{\infty} y(t) \cdot e^{-i\omega' t} dt$$

Substituting y(t) in gives:

$$Z(\omega) = \int_{-\infty}^{\infty} x(t) \cdot \left\{ \frac{1}{2\pi} \int_{-\infty}^{\infty} Y(\omega') \cdot e^{i\omega' t} d\omega' \right\} \cdot e^{-i\omega \cdot t} dt$$

with: ω' an auxiliary angular velocity variable

$$Z(\omega) = \frac{1}{2\pi} \int_{-\infty}^{\infty} \left\{ \int_{-\infty}^{\infty} x(t) \cdot e^{-i(\omega - \omega')t} dt \right\} \cdot Y(\omega') d\omega'$$

$$Z(\omega) = \frac{1}{2\pi} \int_{-\infty}^{\infty} X(\omega - \omega') \cdot Y(\omega') d\omega' \qquad (21)$$

Or in short hand:

$$Z(\omega) = X(\omega) * Y(\omega)$$

This represents the convolution integral in the frequency domain. Signals that are multiplied together in the time domain have their spectra convoluted in the frequency domain.

4.3.5.2 Convolution in Time Domain

Signals that are multiplied together in the frequency domain have their spectra convoluted in the time domain.

Proof: Convolution in time domain is defined by:

$$z(t) = x(t) * y(t) = \int_{-\infty}^{\infty} x(t') \cdot y(t - t') dt' \qquad (22)$$

with: t' an auxiliary time variable

The Fourier Transform of the above is:

$$Z(\omega) = \int_{-\infty}^{\infty} \left\{ \int_{-\infty}^{\infty} x(t') \cdot y(t - t') dt' \right\} e^{-i\omega \cdot t} dt$$

changing the order of integration:

$$Z(\omega) = \int_{-\infty}^{\infty} \left\{ \int_{-\infty}^{\infty} y(t - t') e^{-i\omega \cdot t} dt \right\} x(t') \cdot dt'$$

The expression in curly brackets represent the Fourier Transform of $y(t)$ except for a time delay t'. A time delay my be expressed in the frequency domain by a multiplication with $e^{-i\omega t'}$ (proof: equation (10)).

$$Z(\omega) = \int_{-\infty}^{\infty} \left\{ Y(\omega) \cdot e^{-i\omega \cdot t'} \right\} x(t') \cdot dt'$$

$$Z(\omega) = Y(\omega) \cdot \int_{-\infty}^{\infty} \left\{ x(t') \cdot e^{-i\omega \cdot t'} \right\} \cdot dt'$$

$$Z(\omega) = Y(\omega) \cdot X(\omega) \qquad (23)$$

This completes the proof.

4.3.5.3 Fourier Transform of a rectangular window

A rectangular window is defined by:

$$f(t) = \begin{cases} 1, & -T/2 \le t \le T/2 \\ 0, & \pm\infty < t < \pm T/2 \end{cases} \tag{24}$$

This is mathematically identical with the rectangular pulse (18). The Fourier Transform was:

$$F(f) = T \cdot \frac{\sin(\pi T \cdot f)}{\pi T \cdot f} \tag{25}$$

$$F(f) = T \cdot \operatorname{sinc}(\pi T \cdot f)$$

- The dc value of the spectrum is: $\qquad\qquad F(f=0) = T \tag{26}$
- The cross over frequencies of the spectrum are: $\quad f = \dfrac{n}{T} \qquad n = \pm 1, \pm 2, \dots \tag{27}$
- The frequency response of the rectangular window has an attenuation of -20dB/dec.

4.3.5.4 Fourier Transform of a Hanning Window

A Hanning Window is defined by:

$$y(t) = \begin{cases} \dfrac{1}{2} \cdot \left(1 + \cos\left(\dfrac{2\pi \cdot t}{T}\right) \right), & |t| \le T/2 \\ 0, & |t| > T/2 \end{cases} \tag{28}$$

Substituting in the Fourier Transform, using the complex form for cos gives:

$$Y(\omega) = \frac{1}{2} \int_{-T/2}^{T/2} \left(1 + \frac{e^{i\frac{2\pi \cdot t}{T}} + e^{-i\frac{2\pi \cdot t}{T}}}{2} \right) \cdot e^{-i\omega \cdot t} dt$$

$$Y(\omega) = \frac{1}{2} \left[\frac{e^{-i\omega \cdot t}}{-i\omega} + \frac{e^{i\left(\frac{2\pi}{T}-\omega\right)t}}{2i\left(\frac{2\pi}{T}-\omega\right)} + \frac{e^{-i\left(\frac{2\pi}{T}+\omega\right)t}}{-2i\left(\frac{2\pi}{T}+\omega\right)} \right]_{-T/2}^{T/2}$$

$$Y(\omega) = \frac{1}{2} \left[\frac{e^{-i\omega\frac{T}{2}} - e^{i\omega\frac{T}{2}}}{-i\omega} + \frac{e^{i\left(\pi-\omega\frac{T}{2}\right)} - e^{i\left(\pi+\omega\frac{T}{2}\right)}}{2i\left(\frac{2\pi}{T}-\omega\right)} + \frac{e^{-i\left(\pi+\omega\frac{T}{2}\right)} - e^{-i\left(\pi-\omega\frac{T}{2}\right)}}{-2i\left(\frac{2\pi}{T}+\omega\right)} \right]$$

With $e^{-i\pi} = e^{i\pi} = -1$ follows after some algebraic manipulations:

$$Y(\omega) = \frac{1}{2} \left[\frac{e^{-i\omega\frac{T}{2}} - e^{i\omega\frac{T}{2}}}{-i\omega} + \frac{-2\omega \cdot \left(e^{-i\left(\omega\frac{T}{2}\right)} - e^{i\left(\omega\frac{T}{2}\right)} \right)}{2i\left(\left(\frac{2\pi}{T}\right)^2 - \omega^2 \right)} \right]$$

Upon further simplification:

$$Y(\omega) = \frac{1}{2} \left[\frac{2\left(\frac{2\pi}{T}\right)^2 \cdot \left(e^{i\left(\omega\frac{T}{2}\right)} - e^{-i\left(\omega\frac{T}{2}\right)} \right)}{\omega \cdot \left(\left(\frac{2\pi}{T}\right)^2 - \omega^2 \right) \cdot 2i} \right]$$

$$Y(\omega) = \left(\frac{2\pi}{T}\right)^2 \cdot \frac{\sin\left(\frac{\omega T}{2}\right)}{\omega \cdot \left(\left(\frac{2\pi}{T}\right)^2 - \omega^2\right)} \tag{29}$$

- The first cross over frequencies of the spectrum follows from the second zero of the sin function:

$$\frac{\omega T}{2} = 2 \cdot \pi$$

$$f = \frac{2}{T} \tag{30}$$

- The dc value for the spectrum is found by using L' Hospitals rule:

$$Y(\omega = 0) = \frac{T}{2} \tag{31}$$

- The frequency response of the Hanning window has an attenuation of -60dB/dec.

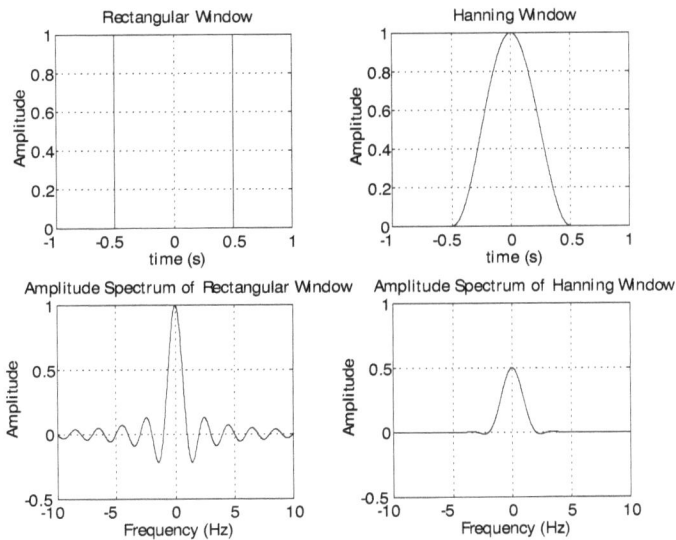

- figure 3:Fourier Transform of Rectangular and Hanning Window (T = 1.0 sec)

4.3.5.5 Periodic Function through Rectangular Window

The periodic function is expressed in exponential form:

$$x(t) = c_n \cdot e^{i 2\pi f_n t}$$

The windowing process was defined by (20):

$$z(t) = x(t) \cdot y(t)$$

where: actual signal: $x(t)$ *periodic function*
        window function: $y(t)$ *rectangular window*
        recorded signal: $z(t)$

Substituting $z(t)$ in the Fourier Transform gives:

13

$$Z(f) = \int_{-T/2}^{T/2} c_n e^{i2\pi f_n t} \cdot e^{-i2\pi f \cdot t} dt$$

$$Z(f) = c_n \int_{-T/2}^{T/2} e^{-i2\pi(f-f_n)t} dt$$

Recognising the similarity with deriving equation (19) it follows immediately:

$$Z(f) = c_n T \cdot \frac{\sin\left(\pi T \cdot (f - f_n)\right)}{\pi T \cdot (f - f_n)} \tag{32}$$

Multiplication of two signals in the time domain is equivalent to convolution of their spectra. Here the spectrum of a periodic signal, which is represented by a single spectral line of amplitude c_n at frequency f_n is convoluted with the spectrum of the rectangular window, which is of form $sin(x)/x$. The resulting spectrum is of weighted $sin(x)/x$ form around frequency f_n of the periodic signal.

The peak value of the spectrum is: $\qquad\qquad Z\left((f - f_n) = 0\right) = c_n T$

The cross over frequency of the spectrum are: $\qquad f - f_n = \dfrac{n}{T} \quad n = \pm 1, \pm 2,.$

Due to convolution of the periodic signal with a rectangular window the line spectrum of the of the periodic signal has changed to a broad peak (was 'smeared') and additional side lobes are present. The side lobes tend to obscure details of the true spectrum.

Assume two periodic functions with frequency f_1 and f_2 that went trough a rectangular window. The resulting Fourier Transform will be the sum of two spectra of the above (32) form (proof: linearity of the FT). If the frequencies are to close the two peaks in the spectrum are obscured and thus can not be distinguished. In order to distinguish two adjacent peaks in the spectrum the first cross over frequency must be smaller than the frequency difference $(f_1 - f_2)$. The minimum frequency difference that can be resolved is called *spectral resolution*.

spectral resolution for rectangular window is: $\qquad f_{sr} = \dfrac{1}{T} \tag{33}$

4.3.5.6 Periodic Function through Hanning Window

The windowing process may be written as (20):
$$z(t) = x(t) \cdot y(t)$$

where: actual signal: $\qquad x(t) \qquad$ *periodic function*
$\qquad\quad$ window function: $\quad y(t) \qquad$ *Hanning window*
$\qquad\quad$ recorded signal: $\quad z(t)$

The periodic function is expressed in exponential form:
$$x(t) = c_n \cdot e^{i2\pi f_n t} = c_n \cdot e^{i\omega_n t}$$

Substituting $z(t)$ in the Fourier Transform gives:

$$Z(\omega) = \frac{1}{2} \int_{-T/2}^{T/2} \left(1 + \frac{e^{i\frac{2\pi \cdot t}{T}} + e^{-i\frac{2\pi \cdot t}{T}}}{2} \right) \cdot c_n e^{i\omega_n t} \cdot e^{-i\omega \cdot t} dt$$

$$Z(\omega) = \frac{c_n}{2} \int_{-T/2}^{T/2} \left(1 + \frac{e^{i\frac{2\pi \cdot t}{T}} + e^{-i\frac{2\pi \cdot t}{T}}}{2} \right) \cdot e^{-i(\omega - \omega_n)t} dt$$

Recognising the similarity to deriving equation (29) it follows immediately:

14

$$Z(\omega) = c_n \left(\frac{2\pi}{T}\right)^2 \cdot \frac{\sin\left(\frac{(\omega - \omega_n)T}{2}\right)}{(\omega - \omega_n) \cdot \left(\left(\frac{2\pi}{T}\right)^2 - (\omega - \omega_n)^2\right)} \tag{34}$$

Once again multiplication of two signals in the time domain is equivalent to convolution of their spectra in the frequency domain. Here the spectrum of a periodic signal, which is represented by a single spectral line of amplitude c_n at frequency f_n is convoluted with the spectrum of the Hanning window. The resulting spectrum is of the form of a weighted Hanning window spectrum around the frequency f_n of the periodic signal.

The first cross over frequencies is:

$$f - f_n = \frac{2}{T}$$

The peak value for the spectrum is:

$$Z\left((f - f_n) = 0\right) = \frac{c_n T}{2}$$

The peak value of the Hanning window spectrum is half the value of the original periodic signal. The peak is two times broader and the side lobes are considerably smaller than for a rectangular window. The result is a poorer resolution of strong peaks but a better definition of the spectral characteristics lying adjacent to a strong peak.

The spectral resolution of a Hanning window is: $f_{sr} = \frac{2}{T}$ (35)

An example plot of an **sin** wave (amplitude 1.0, frequency 4Hz) through a rectangular and a Hanning window (window length 1.0 sec.) and the corresponding Fourier Transform is given in figure 4. The Fourier Transform is obvious using the complex notation for sin and transformation of each term in equation (13) according to equation (32) and (34).

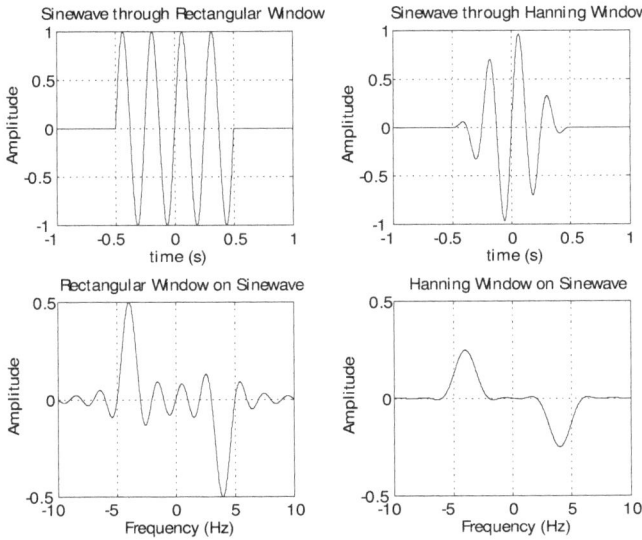

• figure 4: FT of sine wave (f=4.0Hz) through rectangular and Hanning window (T=1.0s)

4.4 Discrete Fourier Transform

4.4.1 Definition of the Discrete Fourier Transform

Sampled data signals have defined values only at certain instances of time, and arise whenever continuos functions are measured or recorded intermittently. Using digital equipment a continuous signal is sampled in the time domain. The sampled signal is discrete and may be expressed mathematically by a series of weighted Dirac functions. Assume the sampling process is perfect and the *sampling interval* τ.

$$f_s(t) = \sum_{n=-\infty}^{\infty} f(n\tau)\cdot\delta(t-n\tau) \qquad n = \text{integer} \tag{36}$$

In practice only a finite *number of samples* N is taken over a finite interval of time, the *observation time* T_0.

$$f_s(t) = \sum_{n=0}^{N-1} f(n\tau)\cdot\delta(t-n\tau) \tag{37}$$

where: sampling interval τ
           number of samples: N

thus: observation time $T_o = N\tau$ (38)

           sampling frequency: $f_s = \frac{1}{\tau}$ (39)

The Discrete Fourier Transform for the sampled signal is given by:

Discrete Fourier Transform (DFT): $F_s(k\omega_0) = \dfrac{1}{N}\sum_{n=0}^{N-1} f(n\tau)\cdot e^{-i\frac{2\pi\cdot n\cdot k}{N}}$ (40)

Inverse Discrete Fourier Transform (IDFT): $f(n\tau) = \sum_{k=0}^{N-1} F_s(k\omega_0)\cdot e^{i\frac{2\pi\cdot n\cdot k}{N}}$ (41)

Note that a discrete signal in the time domain is transformed into a discrete spectrum in the frequency domain. It is important to remember that the discrete spectrum arises from the sampling of the signal in the time domain, e.g. from the use of digital equipment to process the signal. The Fourier Transform of the original underlying continuous signal is by definition continuous.

4.4.2 The DFT spectrum is periodic in Frequency

The discrete spectrum $F_s(k\omega_0)$ of a signal $f(t)$ is periodic with the sampling frequency f_s.

$$\omega_s = \frac{2\pi}{\tau} \qquad \text{or:} \qquad f_s = \frac{1}{\tau}$$

proof: let $k = (k+N)$ in equation (40)

$$F_s\big((k+N)\omega_0\big) = \frac{1}{N}\sum_{n=0}^{N-1} f(n\tau)e^{\frac{-i\cdot 2\pi\cdot n(k+N)}{N}}$$

$$F_s\big((k+N)\omega_0\big) = \frac{1}{N}\sum_{n=0}^{N-1} f(n\tau)e^{\frac{-i\cdot 2\pi\cdot nk}{N}}\cdot e^{i\cdot 2\pi\cdot n}$$

$$F_s\big((k+N)\omega_0\big) = \frac{1}{N}\sum_{n=0}^{N-1} f(n\tau)e^{\frac{-i\cdot 2\pi\cdot nk}{N}}$$

$\therefore$ $F_s\big((k+N)\omega_0\big) = F_s(k\omega_0)$ This completes the proof.

---

[4] P. A. Lynn, page 57
[5] G. Jenkins, Signal Analysis 2, page 2-3

4.4.3 The DFT spectrum is symmetric about the in Nyquist frequency

The spectrum of a Discrete Fourier Transform is symmetric about the Nyquist frequency f_{ny}. (folding frequency) which is half the sampling frequency f_s.

$$f_{ny} = \frac{f_s}{2}$$

proof: let $\quad k = \left(\frac{N}{2} \pm k\right) \quad$ in equation (40)

$$F_s\left(\left(\frac{N}{2} \pm k\right)\omega_0\right) = \frac{1}{N}\sum_{n=0}^{N-1} f(n\tau) e^{\frac{-i \cdot 2\pi \cdot n\left(\frac{N}{2} \pm k\right)}{N}}$$

$$F_s\left(\left(\frac{N}{2} \pm k\right)\omega_0\right) = \frac{1}{N}\sum_{n=0}^{N-1} f(n\tau) e^{\frac{\mp i \cdot 2\pi \cdot nk}{N}} \cdot e^{-i \cdot \pi \cdot n}$$

$$F_s\left(\left(\frac{N}{2} \pm k\right)\omega_0\right) = \frac{1}{N}\sum_{n=0}^{N-1} f(n\tau) e^{\frac{\mp i \cdot 2\pi \cdot nk}{N}} \cdot (-1)^n$$

$$\therefore \quad F_s\left(\left(\frac{N}{2} + k\right)\omega_0\right) = F_s\left(\left(\frac{N}{2} - k\right)\omega_0\right) \qquad \text{This completes the proof.}$$

4.4.4 Practical Considerations

The primary goal to use the DFT is to approximate the Fourier Transform of a continuous time signal. Due to the limitations of the DFT there are three possible phenomena that could result in errors between computed and desired transform.

4.4.4.1 Aliasing

It has been found that the discrete spectrum $F_s(k\omega_0)$ of a signal $f(t)$ is periodic with the sampling frequency f_s and symmetric about the Nyquist frequency f_{ny}. Thus the discrete spectrum of a sampled data signal is repetitive in form.

Assume a continuous spectrum $F(\omega)$ of a signal $f(t)$ with all significant frequency components in the frequency range $-f_{max}$ to $+f_{max}$. The Discrete Fourier Transform of the signal $f(t)$ will be periodic with the sampling frequency f_s and symmetric about the Nyquist frequency f_{ny}. It is evident that if f_{max} is larger than the Nyquist frequency f_{ny} there will be an overlap between adjacent repetitions of the underlying continuous spectrum and therefore the overlap will be 'mirrored' symmetric about the Nyquist frequency. The overlap which arises when the sampling rate is to low is referred to as aliasing.

The minimum sampling frequency $f_{s\ min}$ in order to avoid aliasing is:

$$f_{s\min} \geq 2 \cdot f_{max} \qquad\qquad (42)$$

The sampling theorem may therefore be stated as follows[6]

'A continuous signal which contains no significant frequency components above f_{max}. may in principle be recovered from is sampled version, if the sampling interval is less than $1/2 \cdot f_{max}$ '.

---

[6] P.A. Lynn, page 164

An 'anti aliasing filter' can be used to filter out frequencies above the maximum frequency of interest f_{max}. This is done by using a analogue low pass filter with cut of frequency f_{max} on the input signal prior to sampling.

4.4.4.2 Picket Fence Effect

The DFT is discrete in the time and frequency domain. The computation of the spectrum is limited to integer multiples of the fundamental frequency. Therefore observation of the spectrum with the DFT is analogous to looking through a sort of 'picket fence'. Because the exact behaviour is observed only at discrete points major peaks of the continuous spectrum could lie between two discrete points and the peak might not be detected.

The increment between spectral components can be thought of as the *frequency resolution* f_{fr} and follows immediately from equation (40)

$$f_{fr} = \frac{1}{T_o} = \frac{1}{N \cdot \tau} = \frac{f_s}{N} \qquad (43)$$

4.4.4.3 Leakage

In practice we must limit observation of the signal to a finite interval of time. This is equivalent to multiplying the signal by a rectangular a window function. If the window length of the rectangular window is not a integer multiple of the basic period of the sampled signal a sudden discontinuity is introduced, which causes high frequency terms in the computed spectrum. Additionally a dc value in the spectrum is computed since the average (in the time domain) of the recorded signal is no longer zero. In order to avoid the problem a appropriate window function must be used (e.g. Hanning window).

4.5 Fast Fourier Transformation

The Fast Fourier Transform (FFT) is a computer algorithm devised by Cooley and Tukey in 1965 for efficient computation of the DFT. The number of multiplication's to compute the FFT is $N \cdot \log_2 N$, for DFT: N^2. Thus computation time is reduced significantly and because of fewer round off errors the accuracy is improved.

The number of samples must be a power of 2 for efficient use of the FFT.
$$N = 2^m \qquad m=1,2,3,4, \dots \qquad (44)$$

- table 1: comparsion of number of computations for DFT and FFT

N	DFT (N^2)	FFT ($N \cdot \log_2 N$)	ratio FFT/DFT
16	256	64	0.250
32	1024	160	0.156
64	4096	384	0.0938
128	16384	896	0.0547
256	65536	2048	0.0312
512	262144	4608	0.0176
1024	1048576	10240	0.0098
2048	4194304	22528	0.0054
4096	16777216	49152	0.0029

Refer to literature for more details on the algorithm.

---

[7] P.A. Lynn, page 64

5 Application of the FFT

MATLAB was used to demonstrate the properties of the Fast Fourier Transform. The FFT of known input signals are calculated. For all examples the record length is 1 second and a sampling rate of 1024 Hz was used, resulting a frequency resolution of 1 Hz.

5.1 sin-wave

The input signal is a sin-wave of frequency 50 Hz, amplitude 1.0 and no phase shift. Fifty complete sin waves fall in the sampling interval.

input: sine wave, frequency: $f_1 = 50\ Hz$,   amplitude: $A_1 = 1.0$

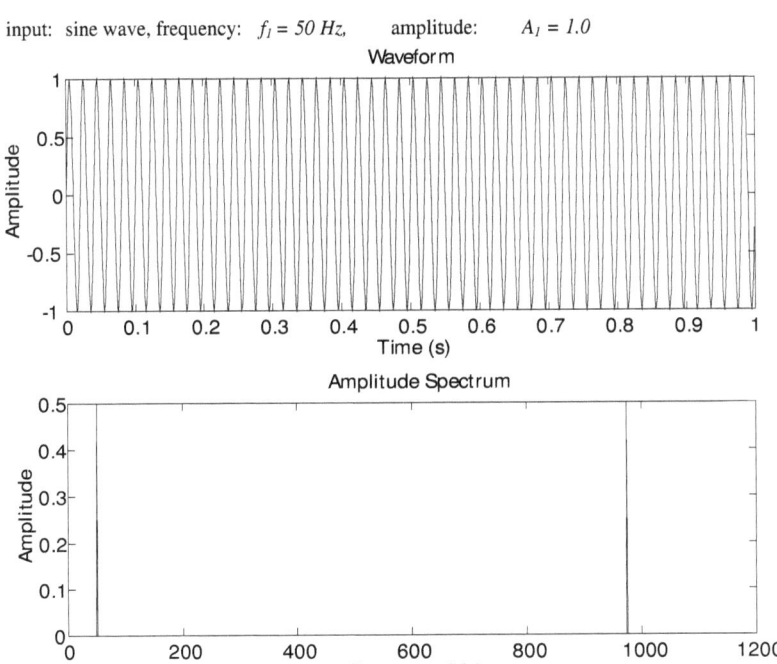

- figure 5: FFT of a sine wave

The theoretical FT of the sin wave (16) has tow discrete spectral lines of amplitude ∓0.5 at frequency ±50 Hz. The result of the FFT is one spectral line of amplitude 0.5 at 50 Hz and one spectral line of amplitude 0.5 at 974 Hz.

The difference between FT and FFT are caused by the properties of the discrete FFT:

- the spectrum is periodic with the sampling frequency: $f_s = 1024\ Hz$

- the spectrum is symmetric about the Nyquist frequency: $f_n = \dfrac{f_s}{2} = 512 Hz$

Thus the spectral line at frequency 50 Hz is repeated symmetric about the Nyquist frequency of 512 Hz resulting in an additional spectral line at 974 Hz or in terms of periodicity the spectral line at -50 Hz is found with the period of 1024 Hz at 974 Hz. The sign of the negative amplitude is lost since absolute values are plotted.

5.2 two superimposed sin-waves

The input signal consists of two superimposed sin-waves. The first sin wave has a frequency of 50 Hz, an amplitude of 1.0 and no phase shift. The second sin wave has a frequency of 700 Hz, an amplitude of 0.5 and no phase shift

input: two superimposed sin waves

       1st sine wave: frequency: $f_1 = 50\ Hz$ amplitude: $A_1 = 1.0$

       2nd sine wave: frequency: $f_2 = 700\ Hz$ amplitude: $A_2 = 0.5$

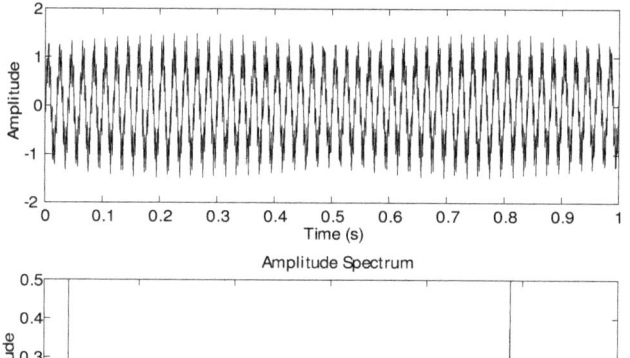

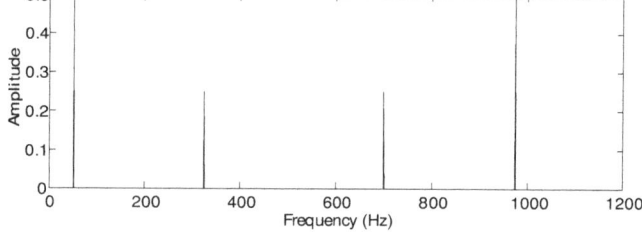

- figure 6: FFT of two superimposed sine wave

The theoretical FT of the superimposed signal are four discrete spectral lines. The first pair of spectral lines are from the fist sine wave with a frequency of 50 Hz and amplitude 1.0. The resulting spectral lines are of amplitude ∓ 0.5 and frequency $\pm 50 Hz$. The second pair of spectral lines are from the second sine wave with a frequency of 700 Hz and amplitude 0.5. The resulting spectral lines are of amplitude ∓ 0.25 and frequency $\pm 700 Hz$. The result of the FFT are four spectral lines. Two spectral line with amplitude 0.5 at frequency 50 Hz and 974 Hz and two spectral lines with amplitude 0.25 at frequency 324 Hz and 700 Hz.

This demonstrates the property of the discrete FFT:

- the spectrum is periodic with the sampling frequency: $f_s = 1024\ Hz$

- the spectrum is symmetric about the Nyquist frequency: $f_n = \dfrac{f_s}{2} = 512 Hz$

- aliasing

The spectral line at 50 Hz is repeated symmetric about the Nyquist frequency 512 Hz resulting in an additional spectral line at 974 Hz. The spectral line of the second signal with an frequency of 700 Hz is found in the FFT spectrum at 700 Hz. Due to the symmetry about the Nyquist frequency of 512 Hz one additional spectral line is found at 324Hz, even though the original signal has no frequency content at 324 Hz! This is referred to as aliasing.

Noteworthy is that if the spectral content of the original input signal were unknown aliasing could lead to misinterpretation of the spectral content of the signal. Parts of the FFT spectrum could be caused by aliasing spectral lines at frequencies beyond the Nyquist frequency. This argument leads into choosing a appropriate sampling frequency and an anti alias filter in order to avoid aliasing.

5.3 two superimposed sin-waves with added random noise

Random noise was added to the signal described in 5.2. Figure 7 - upper right shows the FFT of this signal. Figure 7 - lower left - shows the lower half of the FFT of the signal. Figure 7 - lower right - shows a average spectrum that was obtained by averaging the FFT of four signals with added random noise.

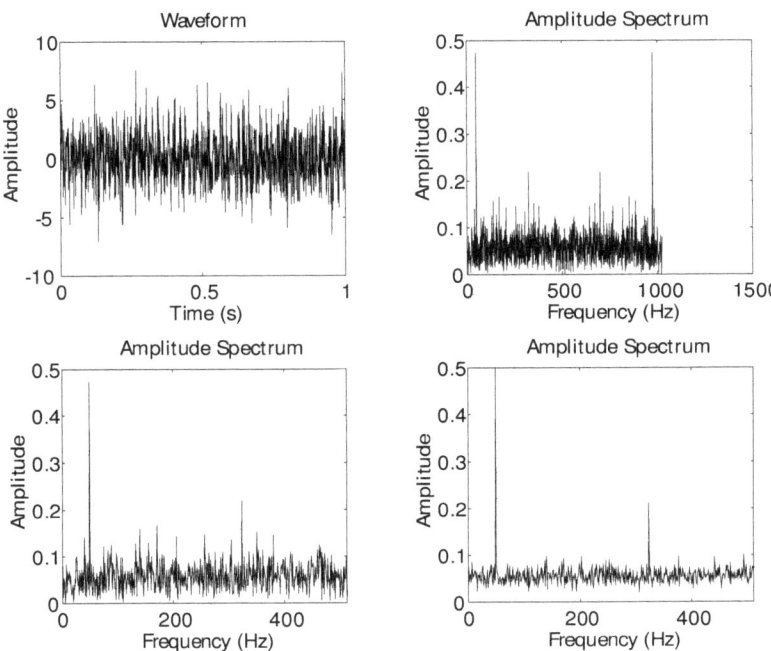

- figure 7: FFT of two superimposed sine waves with added random noise

The mean value of a random signal is zero. The FT of a random signal has randomly distributed spectral lines over the hole frequency domain.

In figure 7 - lower left - the spectral line of the first sin wave at 50 Hz and amplitude 0.5 can be found. The spectral line of the second sin wave with frequency 700 Hz and amplitude 0.25 folded about the Nyquist frequency 512 Hz is expected to be at 324 Hz. This line is found, however the amplitude of the spectral line is not significantly larger than the amplitude of spectral lines caused by the random noise, and therefore could be interpreted as random!

In figure 3 - lower right - the spectral lines of the two sine waves can be identified clearly. The spectral average of the four signals with added random noise averages the random contend of the spectrum, which reduces the spectral lines of the noise significantly. The spectral lines of the two sin waves can be clearly distinguished from the noise.

The effect of random noise is further reduced by averaging a larger number of spectrums.

5.4 square wave

The input signal is a square-wave of frequency 50 Hz, amplitude 1.0 and no phase shift.

input: square wave, frequency: $f_3 = 50\ Hz$ amplitude: $A_3 = 1.0$

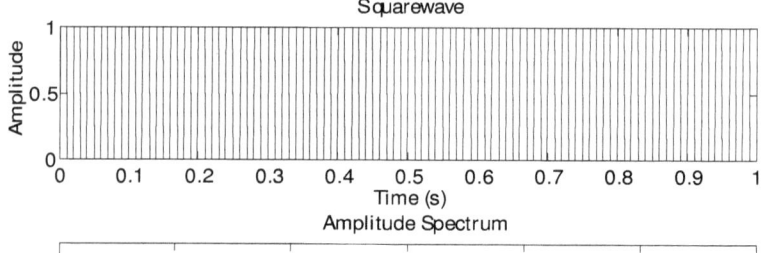

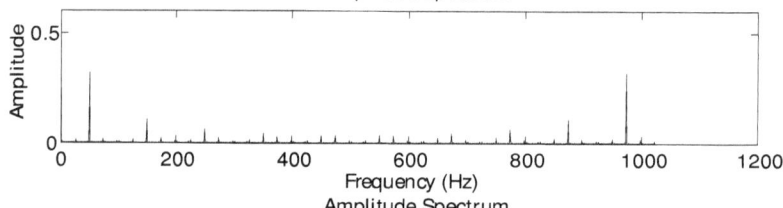

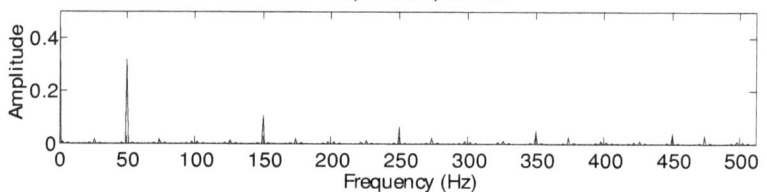

- figure 8: FFT of a square wave

The FT of a square wave(17) has spectral lines only for the odd harmonics, e.g. for 50 Hz, 150 Hz, 250 Hz , ad infinitum. The magnitude of the spectral lines is given by equation (9).
The FFT of the square wave demonstrates again the property of the discrete FFT:

- the spectrum is periodic with the sampling frequency: $f_s = 1024\ Hz$

- the spectrum is symmetric about the Nyquist frequency: $f_n = \dfrac{f_s}{2} = 512 Hz$

- aliasing

The first odd harmonics are captured correctly at 50 Hz, 150 Hz, 250, 350 and 450 Hz. After that point the frequency of the odd harmonics is higher than the Nyquist frequency of 512 Hz, thus aliasing is observed. The 550 Hz component is found at 474 Hz, the 550 Hz harmonic is found at 374 Hz, and so on. The highest harmonic captured is 950 Hz which is found in the spectrum at 74 Hz due to aliasing. Harmonics of higher order, e.g. harmonics that have a higher frequency than the sampling frequency of 1024 Hz are not found in the FFT spectrum.

If the spectral content of the original input signal were unknown aliasing could lead to misinterpretation of the spectral content of the signal. This can be avoided by use of an anti alias filter.

5.5 narrow pulse

The input signal is a rectangular pulse function with amplitude 1.0 in the time interval from 0.1 to 0.2 seconds.

input: rectangular pulse

$$f(t) = \begin{cases} 1, & 0.1 \leq t \leq 0.2 \\ 0, & 0 < t < 0.1; ... 0.2 < t < 1.0 \end{cases}$$

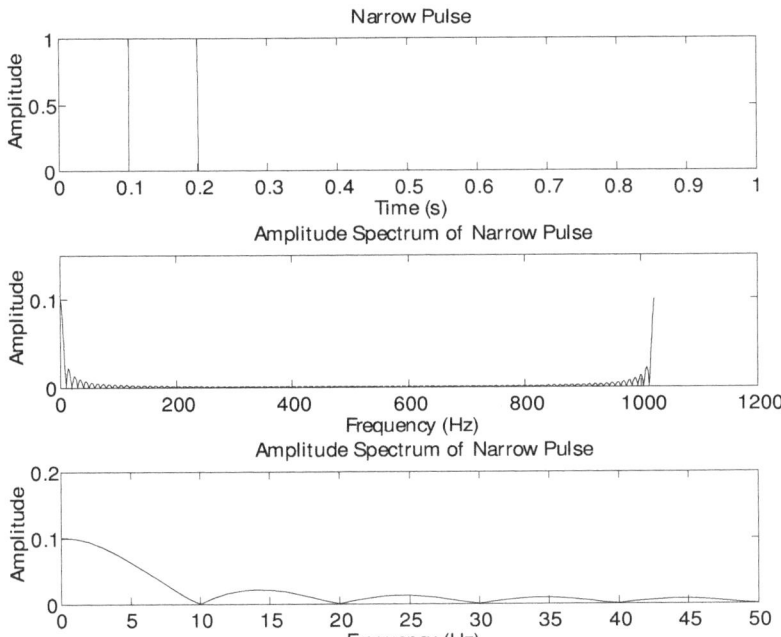

- figure 9: FFT of a narrow pulse

From the Fourier Transform of a rectangular pulse (19) a sinc(x) function with cross over frequency of the spectrum at integer multiples of 10 Hz and a dc value of 0.1 is expected.

The FFT shows the expected characteristic. The frequency resolution of 1 Hz is sufficient to display the sinc(x) function. Negative values of the sinc(x) function are not shown since the absolute values of the spectrum are plotted. Again symmetry about the Nyquist frequency of 512 Hz is observed.

5.6 broad pulse

The input signal is a rectangular pulse function with amplitude 1.0 in the time interval from 0.0 to 0.5 seconds.

input: rectangular pulse

$$f(t) = \begin{cases} 1, & 0.0 \leq t \leq 0.5 \\ 0, & 0.5 < t < 1.0 \end{cases}$$

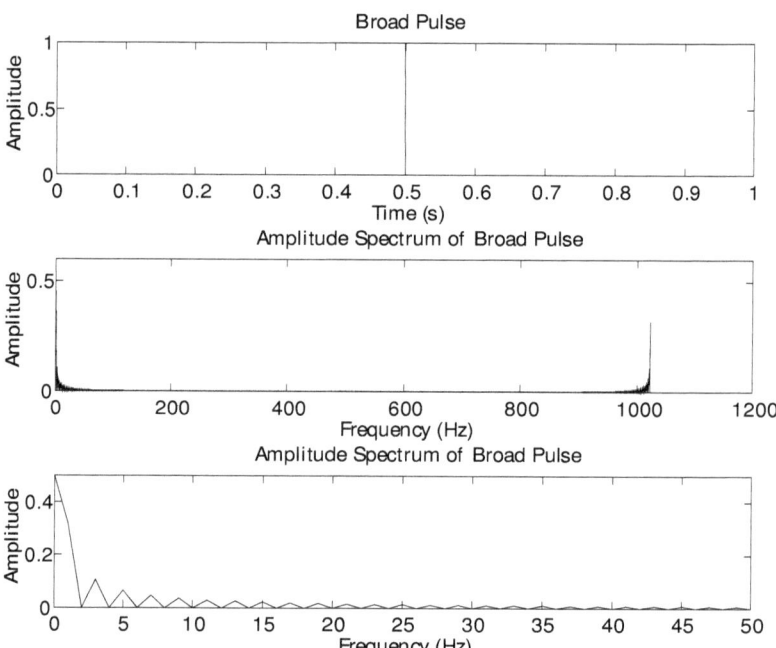

- figure 10: FFT of a broad pulse

From the Fourier Transform of a rectangular pulse (19) a sinc(x) function with cross over frequency of the spectrum at integer multiples of 2 Hz and a dc value of 0.5 is expected.

The FFT shows not the expected characteristic. For the chosen record length of 1.0 seconds the frequency resolution is 1.0 Hz. Thus from the spectrum which has a zero every 2 Hz only the zero and one intermediate value is computed. The characteristic of the sinc(x) function is lost.

Negative values of the sinc(x) function are not shown since the absolute values of the spectrum are plotted. Again symmetry about the Nyquist frequency of 512 Hz is observed.

Only with the knowledge that the spectrum is supposed to have the form of a sinc(x) function one could interpret the result correctly!

5.7 rectangular window

The input function is a rectangular window with a window length of 1.0 seconds.

input: rectangular window

$$f(t) = \begin{cases} 1, & 0.0 \le t \le 1.0 \\ 0, & t > 1.0 \end{cases}$$

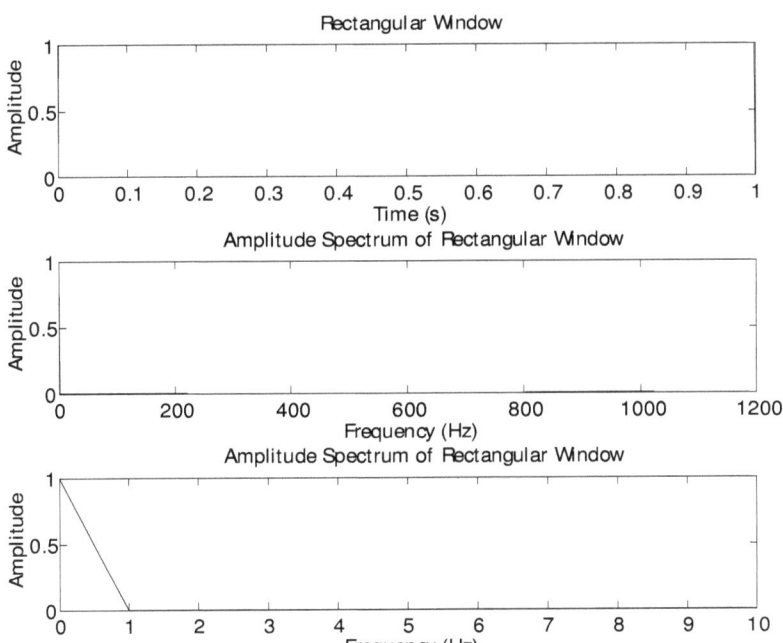

- figure 11: FFT of a rectangular window

From the Fourier Transform of a rectangular window (25) a sinc(x) function with cross over frequency of the spectrum at integer multiples of 1 Hz and an dc value of 1.0 is expected.

The FFT shows not the expected characteristic. For the chosen record length of 1.0 second the frequency resolution is 1.0 Hz. Thus from the spectrum which has a cross over frequency of 1 Hz only the zeros and the dc value of 1.0 are computed. Values of the spectrum at intermediate frequencies are not displayed. This is a good example for the 'picket fence effect'.

5.8 Hanning window

The input function is a Hanning window of window length 1.0 seconds.

input: Hanning window

$$y(t) = \begin{cases} \dfrac{1}{2} \cdot \left(1 + \cos\left(\dfrac{2\pi \cdot (t - 0.5)}{T}\right)\right), & |t| \leq 1.0 \\ 0, & |t| > 1.0 \end{cases}$$

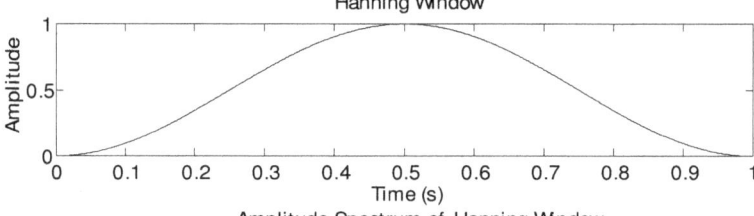

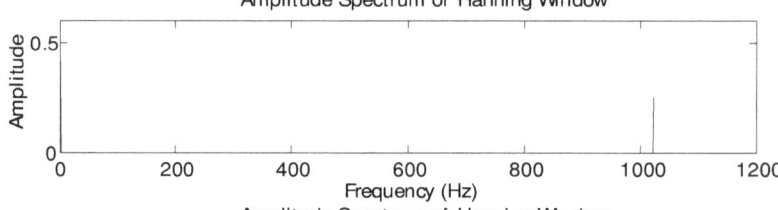

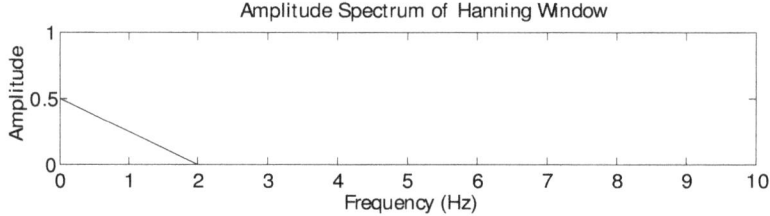

- figure 12: FFT of Hanning window

The Fourier Transform of a Hanning window is given by equation (29). Zeros of the spectrum at integer multiples of 2 Hz and a dc value of 0.5.are expected.

The FFT of a Hanning window shows not the expected characteristic. For the chosen record length of 1.0 second the frequency resolution is 1.0 Hz. Thus from the spectrum which has a cross over frequency of 2 Hz only the zeros, the dc value of 0.5 and one value at 1 Hz are computed. Values of the spectrum at intermediate frequencies are not displayed. This is again a good example for the 'picket fence effect'.

5.9 sine wave trough rectangular and Hanning window

The input signal is a sine-wave of frequency 4 Hz, amplitude 1 amplitude 1.0 and no phase shift. As Window function first a rectangular, than a Hanning window was used.

input: sin wave   frequency: $f_3 = 4\,Hz$ amplitude: $A_3 = 1.0$
        window: 1. Rectangular window (window length 1.0 sec)
                   2. Hanning window   (window length 1.0 sec)

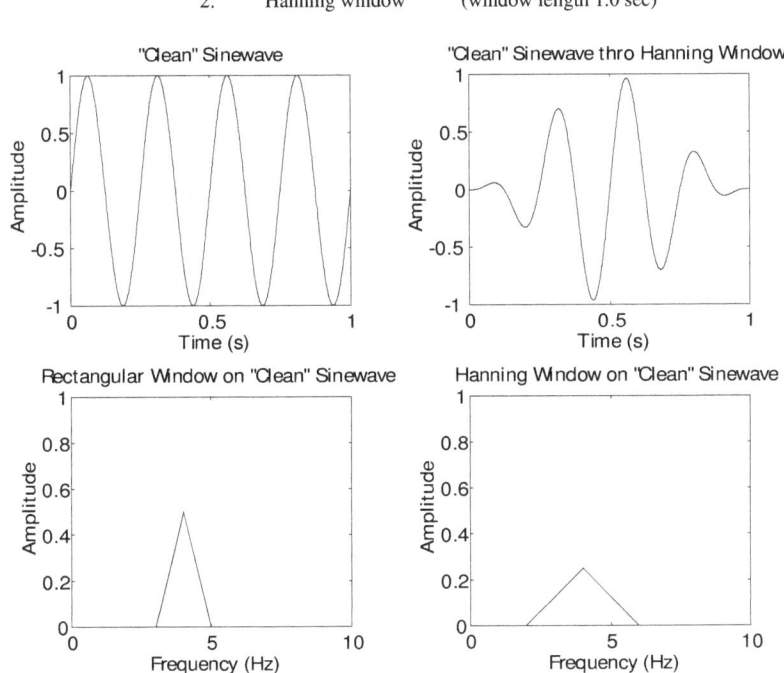

- figure 13: FFT of clean sine wave trough rectangular and Hanning window

From the FT convolution of the sine wave with the window function is expected. The theoretical result is shown in figure 4. For the rectangular window a peak value of amplitude 0.5 at 4.0 Hz is expected. For the Hanning window a peak value of amplitude 0.25 at 4.0 Hz is expected.

For the chosen record length of 1.0 second the frequency resolution is 1.0 Hz. The rectangular window shows only the peak value of 0.5 at 4.0 Hz. All other spectral lines calculated fall at zeros of the spectrum.

The Hanning window shows only the peak value of 0.25 at 4.0 Hz and the intermediate values at 3 and 5 Hz. All other spectral lines calculated fall at zeros of the spectrum.

This is again a good example for the 'picket fence effect'.

5.10 leaky sine wave trough rectangular and Hanning window

The input signal is a sine-wave of frequency 4.25 Hz, amplitude 1.0 and no phase shift. As Window function first a rectangular, than a Hanning window was used.

input: sin wave frequency: $f_3 = 4.25\ Hz$ amplitude: $A_3 = 1.0$
window: 1. Rectangular window (window length 1.0 sec)
2. Hanning window (window length 1.0 sec)

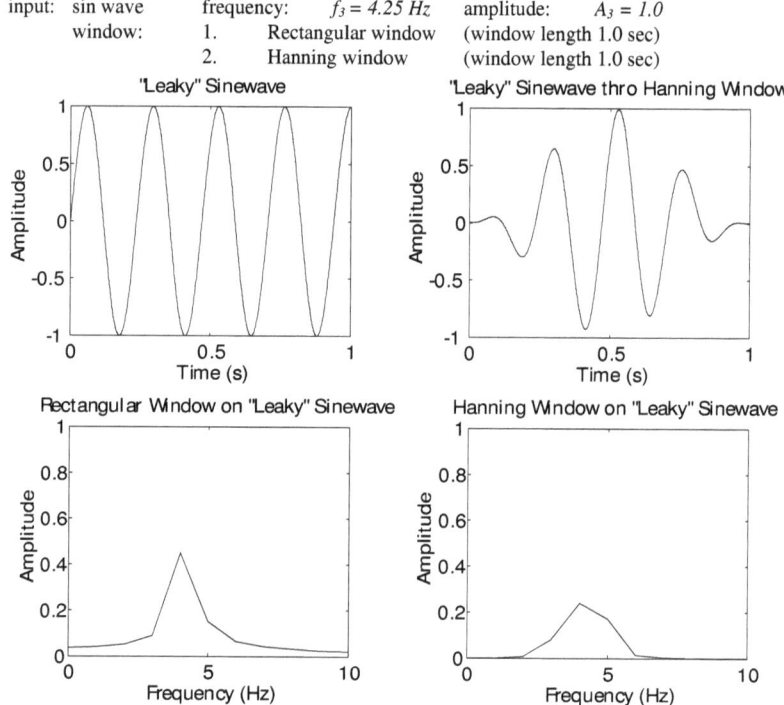

• figure 14: 'leaky' sine wave trough rectangular and Hanning window

From the FT convolution of the sine wave with the window function is expected. For the rectangular window a peak value of 0.5 at 4.25 Hz is expected. .Due to the chosen window length the sin wave is not complete. This will result in a dc value in the spectrum since the time average of the signal is not zero. Furthermore the discontinuity at the end of the signal will cause high frequency content in the spectrum (leakage). For the Hanning window a peak value of 0.25 at 4.24 Hz is expected. The Hanning window has the effect that the signal is forced into a continuous form at the end of the window. Thus no dc value and high frequency content is expected

The FFT using the rectangular window shows dc value and high frequency content of the signal. Due to the frequency resolution of 1 Hz, the peak value of 0.5 at 4.25 Hz is not captured, however a value at 4.0 Hz which is slightly lower than 0.5 is displayed.
The FFT using the Hanning window shows no dc value and high frequency content of the signal. Due to the frequency resolution of 1 Hz, the peak value of 0.25 at 4.25 Hz is not captured, however a value at 4.0 Hz which is slightly lower than 0.5 is displayed.

This is again a good example for the 'picket fence effect'.

5.11 sine wave trough narrow rectangular and Hanning window

The input signal is a sine-wave of frequency 50 Hz amplitude 1.0 and no phase shift. As Window function first a rectangular, than a Hanning window was used. The window length is 0.1 seconds.

input: sin wave frequency: $f = 50 Hz$ amplitude: $A = 1.0$
  window: 1. Rectangular window (window length 0.1 sec)
    2. Hanning window (window length 0.1 sec)

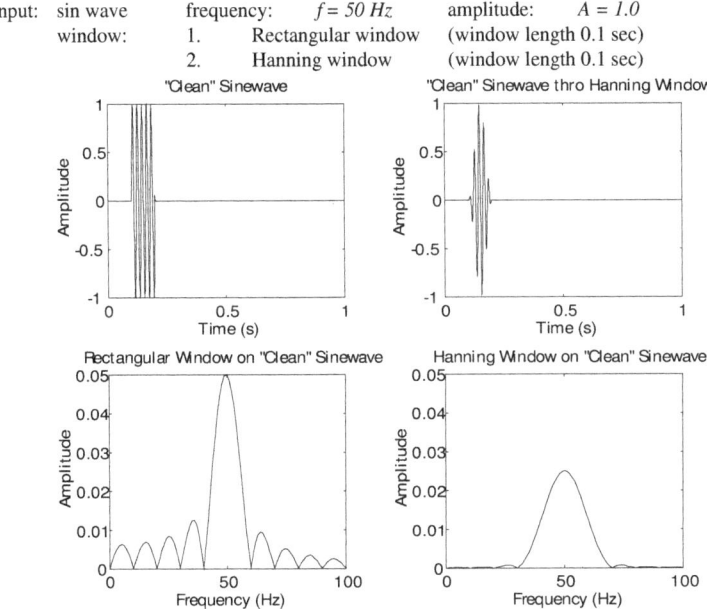

- figure 15: FFT of sine wave through narrow rectangular and Hanning window

From the FT convolution of the sine wave with the window function is expected (figure 4). For the rectangular window a peak value of 0.05 at 50 Hz is expected. For the Hanning window a peak value of 0.025 at 50 Hz is expected.

Convolution of the sine wave with the window function can be observed. For the chosen record length of 1.0 second the frequency resolution is 1.0 Hz. This is sufficient to show the characteristic of the rectangular window, which has a cross over frequency of 10 Hz intervals and Hanning window, which has a cross over frequency of 20 Hz intervals.

The rectangular window displays the correct peak value, however the single spectral line of the input signal at 50 Hz has been 'smeared' to a broad peak and considerable side lobes are present. The Hanning window displays not the correct peak value. The single spectral line of the input signal at 50 Hz has been 'smeared' to a broad peak which has double the width of the peak of the rectangular window. However the side lobes are considerably smaller than for a rectangular window.

Thus the Hanning window does not pick out the main peak, but gives a better resolution for peaks that are close to the main peak.

The width of the peak leads into choosing the correct window length for a desired spectral resolution (equation (33) and (35)).

29

6 Conclusion

The primary goal to use the Discrete Fourier Transform (DFT) is to approximate the Fourier Transform of a continuous time signal. The DFT is discrete in time and frequency domain and has two important properties:
- the DFT is periodic with the sampling frequency
- the DFT is symmetric about the Nyquist frequency

Due to the limitations of the DFT there are three possible phenomena that could result in errors between computed and desired transform.
- Aliasing
- Picket Fence Effect
- Leakage

The DFT of a signal uses only a finite record length of the signal. Thus the input signal for the DFT can be considered as the result of multiplying the signal with a window function. Multiplication in the time domain results in convolution in the frequency domain, which will influence the spectral characteristic of the sampled signal. In the table below rectangular and Hanning window are compared:

		Rectangular Window	Hanning Widow
peak value (input: c_n)	-	$c_n \cdot T$	$c_n \cdot T / 2$
fist zero crossing frequency	[Hz]	$1/T$	$2/T$
attenuation	[dB]	-20	-60
'smearing'	-	small	large
side lobes	-	large	small
leakage	-	yes	no

The Fast Fourier Transform (FFT) is a computationally efficient algorithm for evaluating the DFT of a signal. It is imported to appreciate the properties of the FFT if it is to be used effectively for the analysis of signals. In order to avoid aliasing and resulting misinterpretation of measurement data the following steps should be followed:

1. decide on frequency range of interest, e.g. the bandwidth.
2. match the bandwidth of the anti aliasing filter, e.g. a low pass filter with cut off frequency set to bandwidth
3. choose sampling frequency f_s of analogue digital converter to be 2.5 to 5.0 times the cut of frequency
4. choose window length to give required frequency resolution.
$$T = 1/f_{fr}$$
5. check number of samples $N = T \cdot f_s$, for efficient use of the FFT it is required that:
$$N = 2^m$$
6. repeat step 3, 4 and 5 until an acceptable compromise if found

The effect of random noise can be minimised by averaging several spectra of a signal.

7 Reference

[1] E. Kreysig: Advanced Engineering Mathematics, 5th edition, Jhon Wiley & Sons, 1983

[2] W. Thomson: Theory of Vibration with Application, 2nd edition, Prentice Hall, 1981

[3] P.A. Lynn:An Introduction to the Analysis and Processing of Signals, 3rd edition, Macmillan

[4] E.G. Jenkins: Signals Analysis, Lecture note, Loughborough University, MSc Systems Engineering, Module 1, Engineering Framework

8 Appendix

Appendix 1: Coursework assignment sheet

Appendix 2: MATLAB file signal.m

Appendix 3: some additional MATLAB file used to create various plots

Loughborough University
Department of Aeronautical and Automotive Engineering
and Transport Studies

Ford MSc Course in Automotive Systems Engineering
Mod 1

Engineering Framework

Coursework Assignment 1997
by
Gwyn Jenkins & Stephen Walsh

Digital Signal Processing using the FFT

Objective

     To use MATLAB to demonstrate the properties of the Fast Fourier Transform (FFT) used for the spectral analysis of signals.

Introduction

     Many procedures in the field of vehicle development use sophisticated instrumentation and data acquisition systems to acquire time-histories of physical variables associated with studies of noise-vibration-harshness (NVH), body structural analysis, ride and handling, and various other topics. It is extremely important for the test engineer to understand the (limiting) operational characteristics of the equipment used to ensure acceptable accuracy of the measured data and their subsequent processing.

     It is often more convenient to analyse the data in the frequency-domain rather than in the time-domain for ease of interpretation and understanding. For example, if assessing the effects of measurement, using linear instrumentation with an impulse response of $h(t)$, on an original signal $x(t)$ to obtain a measured time-history $y(t)$ it is necessary to evaluate the convolution integral:

$$y(t) = \int_{0}^{\infty} x(\tau).h(t-\tau)d\tau$$

But it can be shown that *convolution in the time-domain is represented by multiplication in the frequency-domain.*

Giving $Y(s) = H(s).X(s)$ - in terms of the Laplace variable, or $Y(j\omega) = H(j\omega).X(j\omega)$ in terms of the Fourier variable. This leads directly to the very useful concept of a *transfer function (TF)* to describe the dynamic operational characteristics of a linear signal processing element (or system) as: $H(s) = Y(s)/X(s)$, or, $H(j\omega) = Y(j\omega)/X(j\omega)$. It is usual to reresent the TF in terms of a (static) calibration constant (K) and a non-dimensional function of s (or $j\omega$) for dynamic behaviour. Good measurement practice is to **use as much of the measurement range as possible, and to match the bandwidth of the instrumentation system to that of the signal to be analysed.**

When analysing the measured data it is often more informative to study the spectral characteristics than the original time-history, particularly when attempting to identify the cause and/or effect of the measured variable in a physical system. The Fast Fourier Transform (FFT) is a computationally-efficient algorithm for evaluating the Discrete Fourier Transform (DFT) of a time-series (a digitally-sampled time-history) to produce a *discrete* spectrum of the original signal.

The DFT is define as: $F(k\omega_0) = \dfrac{1}{N}\sum_{k=0}^{N-1} f(n\tau).e^{\frac{-j2\pi kn}{N}}$

As with instrumentation, it is necessary to appreciate the properties of the FFT if it is to be used effectively in the analysis of signals.

Procedure

The professional version of MATLAB has a Signal Processing Toolbox containing commands for processing a time-series to give a variety of spectral representations. The most commonly-used of these are available in the student version in the Signals and Systems Toolbox. In this laboratory exercise signals of known characteristics are generated from theoretical functions and used to illustrate the main properties of the FFT method of spectral analysis. The commands are listed on the attached pages.

Report

You are required to submit an *informal* (one page of introduction, one page of conclusions with the main body containing a concise - but comprehensive - *analytical* explanation of the plots obtained during the laboratory session) report to indicate your understanding of the practical application of the FFT in vehicle body engineering.

The report should give an analytical explanation of the following:

1. Convolution in time is represented by multiplication in frequency
2. The FFT spectrum is periodic in frequency
3. The amplitude spectrum is symmetric about the Nyquist frequency
4. The significance of aliasing and how it is minimised
5. Multiplication in time is represented by convolution in frequency.
6. Effects of window functions on spectral resolution and, explanations of the "picket-fence-effect" and "leakage".

Appendix 2: signal.m

```
% *********************************************************
% *                                                       *
% *   Digital Signal Processing using the DFT             *
% *                                                       *
% *   MATLAB code                                         *
% *                                                       *
% *********************************************************
%
% Produce time vector
st=1/1024;t=[0:1023]'*st;
%
% Form sinewave and plot
x1=sin(2*pi*50*t);
subplot(2,1,1);
plot(t,x1);title('Waveform');
xlabel('Time (s)');ylabel('Amplitude');
pause
% Apply FFT and plot amplitude spectrum
x1f=abs(fft(x1))/1024;
hz=[0:1023];
subplot(2,1,2);
plot(hz,x1f);title('Amplitude Spectrum')
ylabel('Amplitude');xlabel('Frequency (Hz)')
pause
%print;
clf
% Add another sinewave at 700 Hz
x2=x1+0.5*sin(2*pi*700*t);
subplot(2,1,1);
plot(t,x2);title('Waveform')
xlabel('Time (s)');ylabel('Amplitude');
pause
% Apply FFT and plot amplitude spectrum
x2f=abs(fft(x2))/1024;
subplot(2,1,2);
plot(hz,x2f);title('Amplitude Spectrum')
ylabel('Amplitude');xlabel('Frequency (Hz)')
pause
%print;
clf
% Add noise and obtain amplitude spectrum
x3=x2+2.0*randn(size(t));
% Adding three other components of noise
% for spectral averaging purposes
x31=x2+2.0*randn(size(t));
x32=x2+2.0*randn(size(t));
x33=x2+2.0*randn(size(t));
% Plot original single waveform analysis
subplot(2,2,1);
plot(t,x3);title('Waveform')
xlabel('Time (s)');ylabel('Amplitude');
pause
% Apply FFT and plot amplitude spectrum
x3f=abs(fft(x3))/1024;
subplot(2,2,2);
plot(hz,x3f);title('Amplitude Spectrum')
ylabel('Amplitude');xlabel('Frequency (Hz)')
pause
% Plot lower half of spectrum
subplot(2,2,3);
plot(hz,x3f);title('Amplitude Spectrum')
ylabel('Amplitude');xlabel('Frequency (Hz)')
axis([0,512,0,0.5]);
pause
% Obtain spectral characteristcs of
% additional components, and average
```

```
x31f=abs(fft(x31))/1024;
x32f=abs(fft(x32))/1024;
x33f=abs(fft(x33))/1024;
x34f=(x3f+x31f+x32f+x33f)/4;
% Plot average spectrum
subplot(2,2,4);
plot(hz,x34f);title('Amplitude Spectrum')
ylabel('Amplitude');xlabel('Frequency (Hz)')
axis([0,512,0,0.5]);
pause
%print;
clf
% Form square wave
for i=1:1024,...
if x1(i)>0,x4(i)=1.0;
else x4(i)=0.0;
end;
end;
subplot(3,1,1);
plot(t,x4);title('Squarewave')
xlabel('Time (s)');ylabel('Amplitude');
pause
% Apply FFT and plot amplitude spectrum
x4f=abs(fft(x4))/1024;
subplot(3,1,2);
plot(hz,x4f);title('Amplitude Spectrum')
ylabel('Amplitude');xlabel('Frequency (Hz)')
pause
% Plot lower half of spectrum
subplot(3,1,3);
plot(hz,x4f);title('Amplitude Spectrum')
ylabel('Amplitude');xlabel('Frequency (Hz)')
axis([0,512,0,0.5]);
pause
%print;
clf
%
%       APERIODIC SIGNALS
%
% Form narrow pulse
t1=[0:102]';t2=[103:205]';t3=[206:1023]';
x51=0*ones(size(t1));x52=ones(size(t2));x53=0*ones(size(t3));
x5=[x51;x52;x53];
subplot(3,1,1);
plot(t,x5);title('Narrow Pulse')
xlabel('Time (s)');ylabel('Amplitude');
pause
% Apply FFT and plot amplitude spectrum
x5f=abs(fft(x5))/1024;
subplot(3,1,2);
plot(hz,x5f);title('Amplitude Spectrum of Narrow Pulse')
ylabel('Amplitude');xlabel('Frequency (Hz)')
pause
% Plot lower part of spectrum
subplot(3,1,3);
plot(hz,x5f);title('Amplitude Spectrum of Narrow Pulse')
ylabel('Amplitude');xlabel('Frequency (Hz)')
axis([0,50,0,0.2]);
pause
%print;
clf
% Form broad pulse
t4=[0:511]';t5=[512:1023]';
x61=ones(size(t4));x62=0*ones(size(t5));
x6=[x61;x62];
subplot(3,1,1);
plot(t,x6);title('Broad Pulse')
xlabel('Time (s)');ylabel('Amplitude');
```

```
pause
% Apply FFT and plot amplitude spectrum
x6f=abs(fft(x6))/1024;
subplot(3,1,2);
plot(hz,x6f);title('Amplitude Spectrum of Broad Pulse')
ylabel('Amplitude');xlabel('Frequency (Hz)')
pause
% Plot lower part of spectrum
subplot(3,1,3);
plot(hz,x6f);title('Amplitude Spectrum of Broad Pulse')
ylabel('Amplitude');xlabel('Frequency (Hz)')
axis([0,50,0,0.5]);
pause
%print;
clf
%
%    Window Functions
%
% Form Rectangular Window
x7=ones(size(t));x7(1)=0.0;x7(1024)=0.0;
subplot(3,1,1);
plot(t,x7);title('Rectangular Window')
xlabel('Time (s)');ylabel('Amplitude');
pause
% Apply FFT and plot amplitude spectrum
x7f=abs(fft(x7))/1024;
subplot(3,1,2);
plot(hz,x7f);title('Amplitude Spectrum of Rectangular Window')
ylabel('Amplitude');xlabel('Frequency (Hz)')
pause
% Plot lower part of spectrum
subplot(3,1,3);
plot(hz,x7f);title('Amplitude Spectrum of Rectangular Window')
ylabel('Amplitude');xlabel('Frequency (Hz)')
axis([0,10,0,1.0]);
pause
%print;
clf
% Form Hanning Window
x8=0.5*(ones(size(t))-cos(2*pi*t));
subplot(3,1,1);
plot(t,x8);title('Hanning Window')
xlabel('Time (s)');ylabel('Amplitude');
pause
% Apply FFT and plot amplitude spectrum
x8f=abs(fft(x8))/1024;
subplot(3,1,2);
plot(hz,x8f);title('Amplitude Spectrum of Hanning Window')
ylabel('Amplitude');xlabel('Frequency (Hz)')
pause
% Plot lower part of spectrum
subplot(3,1,3);
plot(hz,x8f);title('Amplitude Spectrum of Hanning Window')
ylabel('Amplitude');xlabel('Frequency (Hz)')
axis([0,10,0,1.0]);
pause
%print;
clf
%
%  Demonstrate Leakage
%
%
%Form 'clean' sinewave
x9=sin(2*pi*4*t);
subplot(2,1,1);
plot(t,x9);title('"Clean" Sinewave')
xlabel('Time (s)');ylabel('Amplitude');
pause
```

```
% Apply FFT with rect. window and plot amplitude spectrum
x9f=abs(fft(x9))/1024;
% Plot lower part of spectrum
subplot(2,2,3);
plot(hz,x9f);title('Rectangular Window on "Clean" Sinewave')
ylabel('Amplitude');xlabel('Frequency (Hz)')
axis([0,10,0,1.0]);
pause;
% Apply FFT with Hann. window and plot amplitude spectrum
for i=1:1024,...
x10(i)=x9(i)*x8(i);
end;
subplot(2,2,2);
plot(t,x10);title('"Clean" Sinewave thro Hanning Window')
xlabel('Time (s)');ylabel('Amplitude');
pause
% Apply FFT with Hann. window and plot amplitude spectrum
x10f=abs(fft(x10))/1024;
% Plot lower part of spectrum
subplot(2,2,4);
plot(hz,x10f);title('Hanning Window on "Clean" Sinewave')
ylabel('Amplitude');xlabel('Frequency (Hz)')
axis([0,10,0,1.0]);
pause
%print;
clf
% Form 'Leaky' Sinewave
x11=sin(2*pi*4.25*t);
subplot(2,2,1);
plot(t,x11);title('"Leaky" Sinewave')
xlabel('Time (s)');ylabel('Amplitude');
pause
% Apply FFT and plot amplitude spectrum
x11f=abs(fft(x11))/1024;
% Plot lower part of spectrum
subplot(2,2,3);
plot(hz,x11f);title('Rectangular Window on "Leaky" Sinewave')
ylabel('Amplitude');xlabel('Frequency (Hz)')
axis([0,10,0,1.0]);
pause
% Apply FFT with Hann. window and plot amplitude spectrum
for i=1:1024,...
x12(i)=x11(i)*x8(i);
end;
subplot(2,2,2);
plot(t,x12);title('"Leaky" Sinewave thro Hanning Window')
xlabel('Time (s)');ylabel('Amplitude');
pause
% Apply FFT with Hann. window and plot amplitude spectrum
x12f=abs(fft(x12))/1024;
subplot(2,2,4);
plot(hz,x12f);title('Hanning Window on "Leaky" Sinewave')
ylabel('Amplitude');xlabel('Frequency (Hz)')
% Plot lower part of spectrum
axis([0,10,0,1.0]);
pause;
%print;
clf;
```

Appendix 3: winfu01.m

```
% ************************************************************
% *                                                          *
% * Plot spectum of rectangular and Hanning window           *
% *                                                          *
% *                                    A.Kaiser              *
% *                                    April 1997            *
% ************************************************************
%
% Produce frequency vector
st=20/1024;f=[0:1023]'*st-10.0;
%
Tw=1.0;
%
%spectrum of rectangular window
%
for i=1:1024,
 h=pi*f(i)*Tw;
 if h==0
  xfr(i)=Tw;
 else;
  xfr(i)=Tw*sin(h)/h;
 end;
end;
%
%spectrum of hanning window
%
for i=1:1024,
 h=pi*f(i)*Tw;
 c1=Tw*f(i);
 c1=c1*c1;
 c2=1.0/(2.0*pi);
 if h==0
  xfh(i)=0.5*Tw;
 else;
  xfh(i)=((c2*sin(h))/(f(i)*(1.0-c1)));
 end;
end;
%
% Produce time vector
st=2/1024;t=[0:1023]'*st-1.0;
%
% form rectangular window
xtr=cos(2*pi*0.5*t);
for i=1:1024,
 if xtr(i)>0.0,
  xtr(i)=1.0;
 else;
  xtr(i)=0.0;
 end;
end;
% form henning window
for i=1:1024;
 xth(i)=0.0;
end;
for i=257:768,
 xth(i)=0.5*(1.0+cos(2*pi*t(i)));
end;
%
% plot
%
subplot(2,2,1);
plot(t,xtr);title('Rectangular Window');
xlabel('time (s)');ylabel('Amplitude');
axis([-1,1,0.,1.0]);grid;
pause
```

```
subplot(2,2,2);
plot(t,xth);title('Hanning Window')
ylabel('Amplitude');xlabel('time (s)');
axis([-1,1,0.0,1.0]);grid;
pause
subplot(2,2,3);
plot(f,xfr);title('Amplitude Spectrum of Rectangular Window');
xlabel('Frequency (Hz)');ylabel('Amplitude');
axis([-10,10,-0.5,1.0]);grid;
pause
subplot(2,2,4);
plot(f,xfh);title('Amplitude Spectrum of Hanning Window')
ylabel('Amplitude');xlabel('Frequency (Hz)');
axis([-10,10,-0.5,1.0]);grid;
pause
print -dps windfu01.ps;
clf
%
%
%generate sine wave
s1=sin(2*pi*4.0*t)
%
% sine wave through rectangular window
for i=1:1024,
 xtr2(i)=xtr(i)*s1(i);
end;
% sine wave through henning window
for i=1:1024,
 xth2(i)=xth(i)*s1(i);
end;
%
%spectrum of rectangular window shifted 4 Hz
%
for i=1:1024,
 h=pi*(f(i)-4)*Tw;
 hn=pi*(f(i)+4)*Tw;
 if h==0
  xfr2(i)=Tw;
 else;
  xfr2(i)=-0.5*Tw*sin(h)/h;
  xfr2(i)=xfr2(i)+0.5*Tw*sin(hn)/hn;
 end;
end;
%
%spectrum of hanning window shifted 4 Hz
%
for i=1:1024,
 h=pi*(f(i)-4)*Tw;
 c1=Tw*(f(i)-4);
 c1=c1*c1;
 hn=pi*(f(i)+4)*Tw;
 c1n=Tw*(f(i)+4);
 c1n=c1n*c1n;
 c2=1.0/(2.0*pi);
 if h==0
  xfh2(i)=0.5*Tw;
 else;
  xfh2(i)=-0.5*((c2*sin(h))/((f(i)-4)*(1.0-c1)));
  xfh2(i)=xfh2(i)+0.5*((c2*sin(hn))/((f(i)+4)*(1.0-c1n)));
 end;
end;
%
subplot(2,2,1);
plot(t,xtr2);title('Sinewave through Rectangular Window');
xlabel('time (s)');ylabel('Amplitude');
axis([-1,1,-1.0,1.0]);grid;
pause
subplot(2,2,2);
```

```
plot(t,xth2);title('Sinewave through Hanning Window')
ylabel('Amplitude');xlabel('time (s)');
axis([-1,1,-1.0,1.0]);grid;
pause
subplot(2,2,3);
plot(f,xfr2);title('Rectangular Window on Sinewave');
xlabel('Frequency (Hz)');ylabel('Amplitude');
axis([-10,10,-0.5,0.5]);grid;
pause
subplot(2,2,4);
plot(f,xfh2);title('Hanning Window on Sinewave')
ylabel('Amplitude');xlabel('Frequency (Hz)');
axis([-10,10,-0.5,0.5]);grid;
pause
print -dps windsh01.ps;
clf
%
```